C000053502

Printed in Great Britain
by Amazon

35131610R10070